100 *Questions and Answers on* ELECTRICAL SAFETY

Ray A. Jones, PE

JONES AND BARTLETT PUBLISHERS
Sudbury, Massachusetts
BOSTON TORONTO LONDON SINGAPORE

World Headquarters
Jones and Bartlett Publishers
40 Tall Pine Drive
Sudbury, MA 01776
978-443-5000
info@jbpub.com
www.jbpub.com

Jones and Bartlett Publishers Canada
6339 Ormindale Way
Mississauga, Ontario L5V 1J2
Canada

Jones and Bartlett Publishers International
Barb House, Barb Mews
London W6 7PA
United Kingdom

National Fire Protection Association
1 Batterymarch Park
Quincy, MA 02169-7471
www.NFPA.org

Jones and Bartlett's books and products are available through most bookstores and online booksellers. To contact Jones and Bartlett Publishers directly, call 800-832-0034, fax 978-443-8000, or visit our website, *www.jbpub.com.*

Substantial discounts on bulk quantities of Jones and Bartlett's publications are available to corporations, professional associations, and other qualified organizations. For details and specific discount information, contact the special sales department at Jones and Bartlett via the above contact information or send an email to specialsales@jbpub.com.

Production Credits
Publisher—Public Safety Group: Kimberly Brophy
Acquisition Editor—Electrical: Martin Schumacher
Associate Editor: Amanda Brandt
Senior Production Editor: Susan Schultz
Marketing Associate: Meagan Norlund
Manufacturing Buyer: Therese Connell
Composition: Cape Cod Compositors
Cover and Text Design: Anne Spencer
Photo Research Manager/
Photographer: Kimberly Potvin
Photo Researcher: Timothy Renzi
Printing and Binding: Malloy

Library of Congress Cataloging-in-Publication Data
Jones, Ray A., P.E.
100 questions and answers on electrical safety / Ray A. Jones.
p. cm.
Includes index.
ISBN 978-0-7637-5471-6 (pbk.)
1. Electric engineering—Safety measures—Miscellanea. 2. Electric engineering—Insurance requirements—Miscellanea. 3. Electricians—Protection—Miscellanea. I. Title. II. Title: One hundred questions and answers on electrical safety.
TK152.J659 2008
621.319′240289—dc22 2007049974
6048

Printed in the United States of America
12 11 10 09 08 10 9 8 7 6 5 4 3 2 1

Contents

Part 3: Hazard/Risk Analysis 23

Questions 24–27 address hazard and risk analysis, its components, and who must perform the analysis.

Part 4: Personal Protective Equipment 27

Questions 28–40 describe various levels of personal protective equipment and under which circumstances it is required.

Part 7: Safety Equipment 53

Questions 48–55 discuss the use and effectiveness of safety equipment.

Part 8: Arc Flash/Arc Fault 61

Questions 56–70 describe arc flash and arc fault, conditions under which they exist, and how to prevent exposure.

Contents

About the Author

Ray Jones retired from DuPont in 1998, after a 35-year career. During his years at DuPont, Ray was responsible for generating and maintaining corporate and site electrical safety programs across the corporation. He also served as chair of the internal electrical standards program. Since retiring from DuPont, Ray has provided consulting services related to electrical safety programs to many other corporations.

Ray has been a member of the NFPA 70E technical committee for many years and currently serves as chair of that committee. Ray has worked on many national consensus standards, including some published by NFPA, IEEE, and ASTM. He is a member of the Institute of Electrical and Electronics Engineers and has worked on many related committees, conferences, and workshops. Ray is coauthor of five books and 14 technical papers. He guest lectures at various workshops and other venues.

Introduction

100 Questions and Answers on Electrical Safety provides answers to frequently asked questions from practicing electricians, contractors, electrical designers, and inspectors. Although the questions were originally posed to the author to clarify a requirement found in OSHA, NFPA 70E, or other consensus standards, the answers are related to preventing injuries, not just complying with a specific requirement. The answers in this book are based on the writer's 45 years of experience developing, operating, and maintaining electrical safety programs.

In every instance when a question relates to a national standard, you can refer to the developer of the standard for a formal interpretation of the requirement; the discussion in this book is not intended as an interpretation of any written requirement. To prevent injuries and incidents, however, the focus of the effort must be on preventing injuries in general in lieu of complying with a specific written requirement. If you are trying to determine how to comply with regulations, codes, or standards, the following answers may seem inappropriate. If you are trying to prevent injuries or incidents, you should pay attention to the concepts offered in the discussion.

Introduction to Electrical Safety

Why is electrical safety important?

What safety training is required?

More . . .

1 *Why is electrical safety important?*

Several different organizations collect data on injuries in the workplace. Over the last 10 years, on average, every 24 hours a worker was electrocuted on an industrial site from contact with an exposed energized electrical conductor or circuit part, and three other workers were burned by an arcing fault. The same day, another electrical-related fatality occurred in a commercial or residential environment. Electrical installations account for a small part of a facility investment, but electrical injuries account for a major portion of injuries and deaths.

Electrical hazards are not readily visible. Even a trained eye might not identify an electrical hazard. An electrical hazard can be detected only by recognizing and observing indicators. Electricity tends to act like a snake. It is not really dangerous until a person gets too close, and then it strikes. Some snakes are poisonous, and some are not. Similarly, contact with an exposed energized conductor or circuit part can cause a major injury or perhaps cause no injury at all.

Understanding electrical hazards and being able to identify such hazards can reduce the risks associated with electrical installations. Knowledge about electrical safety can help everyone live and work safely around electricity.

2 What is the most important practice to avoid an incident?

Planning is the most effective tool to avoid an incident. Before beginning a task, workers should think about each step necessary to execute the task and identify the expected outcome for each step.

Thus, an effective plan has two parts. For the first part, the worker identifies each step necessary to execute the task. As a necessary aspect of this part of the planning process, the worker should consider what hazards are or might be associated with each step and how he or she might be exposed to them. The second part of an effective plan is to consider what might occur as well as what should occur when the worker performs each step. That way, if the result is different from expectations at a given step, the worker recognizes that he or she must stop the task and generate a new plan.

VOICES OF EXPERIENCE

"All of the sudden there was a huge flash and explosion."

The following illustrates the hidden dangers of working on energized circuits. This happened at a manufacturing plant in Georgia several years ago.

A group of four electricians were working on a panel installing a 400-amp breaker. The panel controlled a large section of the manufacturing plant and it was not feasible to shut down the panel for installation. Everyone was using the proper hot work gloves. Two electricians were wearing glasses and two were not. During the entire removal of the faulty 400-amp breaker and installation of the new breaker, everything was being done in slow motion. All four electricians were communicating with one another about the next move they would be making. I was sitting on top of the panel helping to support the new breaker. All of the sudden there was a huge flash and explosion. The flash was so bright it temporarily blinded us. I received cornea burns from the flash. We also temporarily lost our hearing. The explosion was so loud that the office staff came running to see what had happened, and they were located approximately 300 yards from the panel. The arc shut one half of the plant down at the main entrance.

The plant's electrical department conducted a thorough investigation, which revealed that the contractor had left a bushing off of one leg of the feeder lines for the panel. During a period of less than 1 year the vibration of the plant machinery had worn away the insulation on one leg of a three-phase 2000-amp feeder and it shorted directly to ground.

Fortunately, all of the workers involved in the incident recovered and suffered no long-term effects. This illustrates the danger of working hot even when all safety procedures are followed and everyone is aware of the action going on around them.

Will Alexander
Education Coordinator
Ace Electric, Inc.
Valdosta, Georgia

3 *What safety training is required?*

A better question is what training is necessary. Each employer must provide all necessary information to enable each worker to recognize and avoid electrical hazards. A worker might know how to accomplish a specific work task physically, yet not recognize hazards associated with the task. Various standards, such as NFPA 70E, *Standard for Electrical Safety in the Workplace* and Occupational Safety and Health Administration (OSHA) standards, identify specific information that must be included in the training. It is critical that each employee understand all hazards associated with each task he or she is expected to perform. Employees must understand the hazards associated with their assigned work tasks and how to minimize or avoid the exposure.

Training at each workplace will help workers and supervisors identify the hazards specific to that environment and identify the safety practices utilized at that workplace. Training must be updated whenever there is a change in equipment or in the environment that changes the hazards and risks workers face. Each employer should identify at least one person who is responsible for making sure that safety training reflects the actual hazards and safety practices at the workplace. Similarly, each employer should identify the person or persons responsible for providing safety training, and should track the safety training each worker receives.

How is a person exposed to shock or electrocution?

Electrical current flows between all points that are energized at different voltages. The amount of current depends on the resistance or impedance between the points of different voltage. A person cannot receive a shock or be electrocuted unless he or she contacts an electrical conductor that is both exposed and energized. Touching an exposed energized conductor with a conductive object like a screwdriver is the same as touching the energized conductor with a hand.

Consensus standards identify hazardous voltage at 50 volts. In some instances, 50 volts might not be hazardous, but it should be considered as potentially lethal.

Insulation between a person and an energized conductor reduces or eliminates the risk of shock or electrocution. Air is a good electrical insulator. Rated materials, such as rubber, are also good insulators.

5 *Why is federal and state regulation necessary for what used to be common sense issues?*

Common sense varies widely among knowledgeable and thoughtful people, and common-sense work practices change over time. Technology, equipment, and experience are constantly changing as well. Federal and state regulations are attempts to respond to recognized needs.

Regulations related to safety, the OSHA rules for instance, are a response to widespread experience and statistical evidence that workplace safety was an issue. When the Occupational Safety and Health Act was written in 1970, many thousands of workers were injured on the job each year. The incidents and injuries caused pain and suffering, as well as a reduction in the gross national product from the lost productivity in many workplaces. OSHA rules and similar standards take the common-sense knowledge and practices that exist in the safest workplaces and make those standards applicable to every workplace.

NFPA 70E and similar voluntary standards are attempts by society to support and enhance the regulations. For example, when 70E-1995 was published, electrocution was the fourth leading cause of industrial fatalities. Arc flash injuries were not on the map, even though experience shows that up to 80 percent of electrical injuries were thermal burns from exposure to an arcing fault (Jones, *Electrical Safety in the Workplace*, 2000). Standards and regulations help workers and employers identify risks and hazards, like the risk of thermal burns from an arcing fault, and institute practices that reduce those risks.

As the size of a company increases, the chance that a company's leaders will not know about or understand the risks that workers face also increases. The leaders may have little or no experience with the day-to-day tasks of workers as the workers construct, operate, and maintain corporate facilities—and leaders similarly have little or no experience with the hazards and risks that accompany workers' tasks. Regulations help employers and corporate owners provide safe workplaces, and regulations also provide workers with guidance and protection.

6 *Is lockout/tagout enough to eliminate exposure to electrical energy?*

Lockout/tagout (LOTO) is not enough to eliminate exposure to electrical energy. LOTO is one of six steps necessary to eliminate all potential exposure to electrical energy. If the energy source of potential injury is mechanical, the hazard is likely to be quite visible. However, electrical energy is not visible. Its presence can be determined only by observing clues about its existence. Frequently, electrical energy can reappear. It is necessary to create an electrically safe work condition before all potential exposure has been eliminated.

7 *Is a journeyman electrician a qualified person?*

A journeyman electrician might or might not be a qualified person. To be qualified, a worker must understand how to recognize and avoid exposure to electrical hazards. A worker might be proficient at accomplishing the physical aspects of a work task without understanding the associated hazards, which would make him or her unqualified. A worker could be qualified for one task and unqualified for another task.

8 *Why is coordination among contractors important?*

OSHA requires that each employer have an electrical safety program. The program must contain a procedure that covers lockout/tagout. Employees of each organization must understand how work tasks he or she might perform may increase exposure or risk for an employee of another organization. Many injuries occur when a contractor's employee is exposed to a hazardous condition that an employee of a different contractor creates or allows to exist. For instance, when an electrical contractor installs and energizes a feeder to a new motor control center that supplies existing equipment, employees of the owner may be exposed to electrical hazards. All employers, including owners and contractors, must ensure that employees of other contractors are aware of any unique characteristics or procedures. For large or complex jobs, it is useful to designate a person in charge and create specific procedures that increase communication and reduce these risks.

9 *How do I predict incident energy?*

As Annex D of NFPA 70E notes, several methods are available to estimate incident energy. All available methods provide only estimated results. The Institute of Electrical and Electronics Engineers, Inc. (IEEE) and National Fire Protection Association (NFPA) have a joint research effort underway to produce a recommended method of predicting hazards associated with an arcing fault, including thermal energy.

10 *What is the hazard associated with a live part?*

As currently defined in NFPA 70, *National Electrical Code*®, NFPA 70E, and other national consensus codes and standards, a hazard is not necessarily associated with the term *live part*. If the live part is exposed (uninsulated), all recognized hazards are present, including shock, electrocution, arc flash, and arc blast. An arc flash hazard can be present in some instances, such as through ventilation openings, even with the door closed and latched. In theory, no hazards exist when the live part is insulated.

11 *How do I justify work on or near live parts?*

Work on or near live parts exposes a worker to an elevated risk of injury. However, in some instances, disconnecting the source of electrical energy increases the risk of injury to other people, such as in a hospital operating room. For such work to be justified, the risk associated with removing power must exceed the risk of performing the work on or near energized electrical conductors.

12 *Does a third-party label provide assurance that the equipment is safe?*

Third-party testing ensures that the equipment complies with the conditions defined by the testing procedure. The third-party label ensures that the equipment is safe to operate under the conditions defined by the test procedure. Generally, however, equipment is not tested in arcing fault conditions. The equipment also must be installed and used in accordance with the listing or labeling instructions and any applicable manufacturer's instructions. The third party label does not mean that the equipment is safe under all conditions. Instead, the label means that the equipment is safe to operate if installed correctly.

13 *What is the responsibility of an owner?*

The OSH Act requires employers to provide a safe workplace for employees. Generally, consensus standards assign the same responsibility. Logically, employers are responsible for the environment in which employees work, and if an employee is injured, the employer has some responsibility for the incident.

Owners are in a slightly different category, although an owner also might be the employer. Where contractors are involved with the work, the role of the owner shifts to ensuring that the contractor is aware of all hazards associated with the work. The owner should determine and evaluate each contractor's record and safety program.

Multi-employer circumstances are complex. However, both employers and owners should take steps to identify and discuss hazards, and to avoid or mitigate employee exposure to hazards.

What is the responsibility of an employee?

Consensus standards indicate that employees are responsible for implementing the requirements of the employer's electrical safety program. However, the responsibility goes beyond that simple direction. To achieve maximum success, employees must accept and become involved in the electrical safety program.

What part does maintenance play in controlling exposure to an electrical hazard?

When an installation is complete and the equipment placed into service, it begins to deteriorate. Adequate maintenance extends the life of the equipment and limits unnecessary exposure to unsafe conditions. Some national consensus standards, such as NFPA 70B, *Recommended Practice for Electrical Equipment Maintenance*, provide maintenance recommendations. However, equipment manufacturers are likely the best source of maintenance information.

Safety Standards
(NFPA 70E, OSHA, and Electrical Safety Programs)

What is an electrical safety program?

Who enforces NFPA 70E?

More . . .

16 *What is an electrical safety program?*

An electrical safety program is a concrete manifestation of a company's philosophy about the safety of its workers. It is a set of documents that guide actions and reactions for an employer. The documents define policies, procedures, training, and auditing.

An electrical safety program should develop a standard way to communicate by defining all terms used in that workplace. For example, the term *hot work* is often used for two different meanings: It might refer to work done on an energized circuit or to work that is thermally hot. The term has no consensus meaning. If a workplace uses that term, then the workplace's electrical safety program should define how that term is used.

A specific person or persons should be responsible for maintaining and updating a company's electrical safety program. The person or persons responsible should review and revise the program regularly, taking into account new equipment and machinery, new procedures, and new employees as necessary.

17 *What is the role of an employee in developing an electrical safety program?*

Both OSHA and consensus standards assign the responsibility of providing an electrical safety program to employers. The same documents assign the responsibility of implementing defined requirements to employees. However, for the program to be effective, employees should participate in developing procedures and practices. Employees have hands-on experience with particular equipment, and are frequently familiar with exposure, or of ways of reducing exposure. Electrical safety programs reach maximum effectiveness when employees and employers work together to develop the content of the program.

18 *Why is auditing important?*

Auditing provides information about the condition of the electrical safety program. An effective audit will indicate the state of training and whether defined requirements are adequate. Audits provide information to employers and to employees. After the electrical safety program is in place, auditing is the only viable method to determine if the program is effective without monitoring injuries or incidents.

19 *Who writes NFPA 70E?*

Members of the public write NFPA 70E. The process for generating the standard is guided by the *NFPA Regulations Governing Committee Projects*, available on the NFPA Web site (*www.nfpa.org*). The *NFPA Regulations Governing Committee Projects* meets or exceeds requirements defined by the American National Standards Institute (ANSI). One characteristic of the process is that a technical committee of people expert in the subject being considered discusses public proposals and comments submitted by members of the public. As described in the *NFPA Regulations Governing Committee Projects*, the technical committee includes representatives from all sectors of general industry, chosen through an application process.

Does OSHA enforce NFPA 70E?

No. OSHA enforces its own standards. In the United States, only the U.S. Department of Energy requires compliance with NFPA 70E. Otherwise, NFPA 70E is an American National Standard. As such, the standard describes normal and reasonable measures to prevent injury from electrical hazards. NFPA 70E is a voluntary standard for everyone except employees and contractors under the jurisdiction of the U.S. Department of Energy.

Who enforces NFPA 70E?

As noted, NFPA 70E is an American National Standard. NFPA 70E is not enforced until an organization elects to enforce it. If an organization adopts NFPA 70E, the adopting organization becomes the enforcing agency. Note, however, that NFPA 70E defines reasonable and normal protective measures, rather than exceptional or unusual protective measures.

22 *Does NFPA 70E have an economic benefit?*

An electrical safety program is a sound economic investment. Studies conducted by the Construction Industry Institute report an annual return on investment of between four and eight times. In addition, common sense tells us that preventing injuries means preventing incidents that result in damaged equipment, lost production, medical expenses, additional training expenses, legal expenses, insurance expenses, and similar negative results. NFPA 70E discusses requirements that prevent injuries and associated incidents, which means that following NFPA 70E does have an economic benefit.

What is so important about the notes that follow the tables in consensus standards?

A table establishes a default condition for a very complex process. The limiting characteristics of the complex process are defined by the information contained in the notes. If the limit defined in the notes is exceeded, the table does not apply. For example, Note 1 to NFPA 70E, Table 130.7(C)(9)(a) indicates short circuit current of 25 kA and a clearing time of 2 cycles. If the short circuit current exceeds 25 kA or the clearing time exceeds 2 cycles, the table does not apply and a flash hazard analysis is necessary.

Knowing the limiting conditions of the complex processes means knowing when additional information or analysis is necessary. The table notes tell what information is needed for the default condition to apply. If the situation is different, the default information should not be used.

Hazard/Risk Analysis

What is a hazard/risk analysis?

What are the components of a hazard analysis?

More . . .

24 *What is a hazard/risk analysis?*

A hazard/risk analysis is a process in which a specific work task is considered and all hazards are identified. The work task is analyzed to determine the likelihood that an incident will occur. Workers must then assess the risk of an incident associated with each hazard and determine if the risk might result in an injury. The worker then must determine if the risk of injury is sufficiently low to be acceptable.

25 *Why is risk included in the analysis?*

No injury is possible from a hazard unless a person is exposed to it. Risk is associated with the chance that an incident will occur and, subsequently, the chance that an injury will result from the incident. Some risk is associated with every task. The idea, then, is to determine if the risk of incident and the risk of injury in a particular situation are low enough to be acceptable.

For example, as a worker approaches an exposed energized electrical conductor, the likelihood of contacting the conductor increases. At some approach distance, the likelihood of contact with the exposed energized conductor becomes the primary issue. National consensus standards define approach boundaries to assist workers as they consider the risks associated with approaching an exposed energized electrical conductor.

Keep in mind that two kinds of risk are involved: the risk that an incident will occur, and the risk that an injury will result. A high degree of risk for an incident may exist, while potential for serious injury may be low. Alternatively, only a slight risk of an incident might exist, but there may be a near certainty that a serious injury will result from such an incident.

What are the components of a hazard analysis?

A complete hazard analysis includes identifying all hazards associated with the task. Available hazards might be restricted to electrical hazards, but most work tasks also have nonelectrical hazards associated with them. Falls and falling objects are principal additional hazards. A hazard analysis consists of asking questions either mentally or verbally about whether a hazard exists. For instance, if the work task is not elevated, a fall hazard is unlikely. If one conductor within the enclosure is energized, a shock hazard exists. If a circuit remains energized, the worker must consider the possibility of arc flash.

27 *Who must perform a hazard/risk analysis?*

Workers are exposed to hazards. Therefore, workers are the last in a series of people (beginning with the safety manager and immediate supervisor) who should evaluate hazards and assess the risk of exposure to workers. Workers must be provided with the necessary information to enable a hazard/risk analysis, although supervision should be involved in the analysis process.

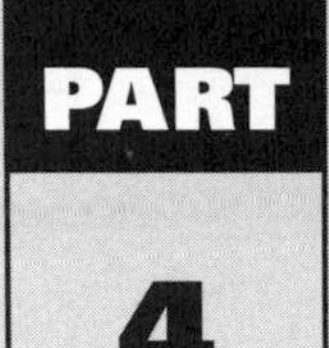

Personal Protective Equipment

What personal protective equipment protects a person from shock or electrocution?

When is a face shield satisfactory face protection?

More . . .

What personal protective equipment (PPE) protects a person from shock or electrocution?

For shock or electrocution to occur, a worker must experience current flow through his or her body. Insulating materials decrease the amount of current. Generally, rubber products are good insulators (Figure 1). In

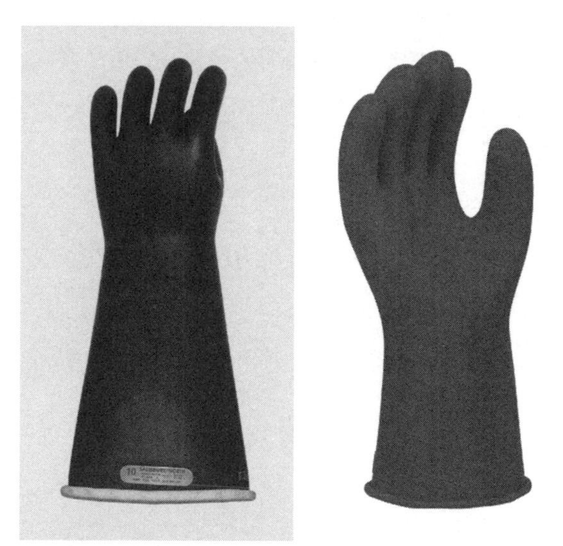

Figure 1 Voltage-rated rubber gloves.
Courtesy of Salisbury Electrical Safety, LLC.

some cases, however, a rubber product might contain a contaminant, be semi-conductive, or intentionally contain a conductive component. Workers should use only insulating products that are assigned a rating by the manufacturer.

Why must conductive apparel such as rings be removed when working on or near live parts?

Conductive apparel provides a mechanism to initiate an arcing fault by creating a short circuit. Conductive apparel may be the low-resistance point of contact that results in an electrocution. Metal components of apparel also absorb and retain thermal energy from an arcing fault. If the metal component is in a worker's pocket, the protective nature of the flame-resistant PPE is decreased.

What is FR clothing?

Flame-resistant or FR clothing protects the wearer from the thermal effects of an arcing fault. FR clothing should be arc rated. FR clothing used for protection from an arcing fault must be rated for use in an environment influenced by an electrical arc. Such apparel is assigned an arc rating in calories per square centimeter (Figure 2).

Some children's clothing and some bedding are rated as flame retardant. Generally however, the flame retardant property is the result of treating the fabric with a chemical. The chemical (and flame retardant property) will deteriorate as the product is laundered.

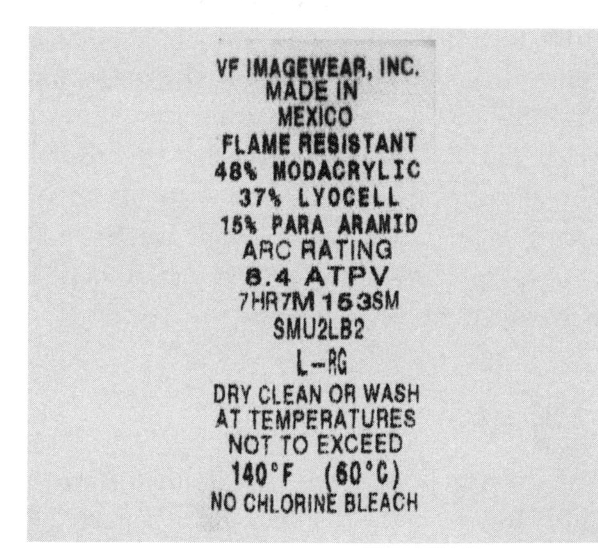

Figure 2 Clothing label.
Product provided courtesy of VF Imagewear, Inc.

1 *When can I use the table method to determine FR protective equipment?*

Table 130.7(C)(9)(a) in NFPA 70E-2004 identifies various work tasks, categorized by equipment type. Using this table is an acceptable method to determine PPE requirements. However, the notes that follow the table (Table 1) describe the conditions under which the table applies.

2 *Why is FR clothing rated in calories per square centimeter?*

All current methods of estimating the thermal hazard in electrical equipment determine incident energy in calories per square centimeter. The unit of measure is not important except that it becomes a standard designation. Arc-rated clothing is rated as calories per square centimeter to enable direct comparison of incident energy and the protective nature of the clothing.

Table 1 Hazard/Risk Category Classifications (NFPA 70E Table 130.7(C)(9)(a))

Task (Assumes Equipment Is Energized, and Work Is Done Within the Flash Protection Boundary)	Hazard/Risk Category	V-rated Gloves	V-rated Tools
Panelboards Rated 240 V and Below—Notes 1 and 3			
Circuit breaker (CB) or fused switch operation with covers on	0	N	N
CB or fused switch operation with covers off	0	N	N
Work on energized parts, including voltage testing	1	Y	Y
Remove/install CBs or fused switches	1	Y	Y
Removal of bolted covers (to expose bare, energized parts)	1	N	N
Opening hinged covers (to expose bare, energized parts)	0	N	N
Panelboards or Switchboards Rated >240 V and up to 600 V (with molded case or insulated circuit breakers)—Notes 1 and 3			
CB or fused switch operation with covers on	0	N	N
CB or fused switch operation with covers off	1	N	N
Work on energized parts, including voltage testing	2*	Y	Y

600 V Class Motor Control Centers (MCCs)— Notes 2 (except as indicated) and 3			
CB or fused switch or starter operation with enclosure doors closed	0	N	N
Reading a panel meter while operating a meter switch	0	N	N
CB or fused switch or starter operation with enclosure doors open	1	N	N
Work on energized parts, including voltage testing	2*	Y	Y
Work on control circuits with energized parts 120 V or below, exposed	0	Y	Y
Work on control circuits with energized parts > 120 V, exposed	2*	Y	Y
Insertion or removal of individual starter "buckets" from MCC—Note 4	3	Y	N
Application of safety grounds, after voltage test	2*	Y	N
Removal of bolted covers (to expose bare, energized parts)	2*	N	N
Opening hinged covers (to expose bare, energized parts)	1	N	N

(Continued)

Table 1 Hazard/Risk Category Classifications (NFPA 70E Table 130.7(C)(9)(a)) ***Continued***

Task (Assumes Equipment Is Energized, and Work Is Done Within the Flash Protection Boundary)	Hazard/Risk Category	V-rated Gloves	V-rated Tools
600 V Class Switchgear (with power circuit breakers or fused switches)—Notes 5 and 6			
CB or fused switch operation with enclosure doors closed	0	N	N
Reading a panel meter while operating a meter switch	0	N	N
CB or fused switch operation with enclosure doors open	1	N	N
Work on energized parts, including voltage testing	2*	Y	Y
Work on control circuits with energized parts 120 V or below, exposed	0	Y	Y
Work on control circuits with energized parts > 120 V, exposed	2*	Y	Y
Insertion or removal (racking) of CBs from cubicles, doors open	3	N	N
Insertion or removal (racking) of CBs from cubicles, doors closed	2	N	N
Application of safety grounds, after voltage test	2*	Y	N
Removal of bolted covers (to expose bare, energized parts)	3	N	N
Opening hinged covers (to expose bare, energized parts)	2	N	N

Other 600 V Class (277 V through 600 V, nominal) Equipment—Note 3			
Lighting or small power transformers (600 V, maximum)	—	—	—
Removal of bolted covers (to expose bare, energized parts)	2*	N	N
Opening hinged covers (to expose bare, energized parts)	1	N	N
Work on energized parts, including voltage testing	2*	Y	Y
Application of safety grounds, after voltage test	2*	Y	N
Revenue meters (kW-hour, at primary voltage and current)	—	—	—
Insertion or removal	2*	Y	N
Cable trough or tray cover removal or installation	1	N	N
Miscellaneous equipment cover removal or installation	1	N	N
Work on energized parts, including voltage testing	2*	Y	Y
Application of safety grounds, after voltage test	2*	Y	N
NEMA E2 (fused contactor) Motor Starters, 2.3 kV Through 7.2 kV			
Contactor operation with enclosure doors closed	0	N	N
Reading a panel meter while operating a meter switch	0	N	N

(Continued)

Table 1 Hazard/Risk Category Classifications (NFPA 70E Table 130.7(C)(9)(a)) ***Continued***

Task (Assumes Equipment Is Energized, and Work Is Done Within the Flash Protection Boundary)	Hazard/Risk Category	V-rated Gloves	V-rated Tools
Contactor operation with enclosure doors open	2*	N	N
Work on energized parts, including voltage testing	3	Y	Y
Work on control circuits with energized parts 120 V or below, exposed	0	Y	Y
Work on control circuits with energized parts > 120 V, exposed	3	Y	Y
Insertion or removal (racking) of starters from cubicles, doors open	3	N	N
Insertion or removal (racking) of starters from cubicles, doors closed	2	N	N
Application of safety grounds, after voltage test	3	Y	N
Removal of bolted covers (to expose bare, energized parts)	4	N	N
Opening hinged covers (to expose bare, energized parts)	3	N	N
Metal Clad Switchgear, 1 kV and Above			
CB or fused switch operation with enclosure doors closed	2	N	N
Reading a panel meter while operating a meter switch	0	N	N
CB or fused switch operation with enclosure doors open	4	N	N

Work on energized parts, including voltage testing	4	Y	Y
Work on control circuits with energized parts 120 V or below, exposed	2	Y	Y
Work on control circuits with energized parts > 120 V, exposed	4	Y	Y
Insertion or removal (racking) of CBs from cubicles, doors open	4	N	N
Insertion or removal (racking) of CBs from cubicles, doors closed	2	N	N
Application of safety grounds, after voltage test	4	Y	N
Removal of bolted covers (to expose bare, energized parts)	4	N	N
Opening hinged covers (to expose bare, energized parts)	3	N	N
Opening voltage transformer or control power transformer compartments	4	N	N
Other Equipment 1 kV and Above			
Metal clad load interrupter switches, fused or unfused	—	—	—
Switch operation, doors closed	2	N	N
Work on energized parts, including voltage testing	4	Y	Y
Removal of bolted covers (to expose bare, energized parts)	4	N	N

(Continued)

Table 1 Hazard/Risk Category Classifications (NFPA 70E Table 130.7(C)(9)(a)) ***Continued***

Task (Assumes Equipment Is Energized, and Work Is Done Within the Flash Protection Boundary)	Hazard/Risk Category	V-rated Gloves	V-rated Tools
Opening hinged covers (to expose bare, energized parts)	3	N	N
Outdoor disconnect switch operation (hookstick operated)	3	Y	Y
Outdoor disconnect switch operation (gang-operated, from grade)	2	N	N
Insulated cable examination, in manhole or other confined space	4	Y	N
Insulated cable examination, in open area	2	Y	N

Notes:
V-rated Gloves are gloves rated and tested for the maximum line-to-line voltage upon which work will be done.
V-rated Tools are tools rated and tested for the maximum line-to-line voltage upon which work will be done.
2* means that a double-layer switching hood and hearing protection are required for this task in addition to the other Hazard/Risk Category 2 requirements of Table 130.7(C)(10).
Y = yes (required) N = no (not required)
1. 25 kA short circuit available, 0.03 second (2 cycle) fault clearing time.
2. 65 kA short circuit available, 0.03 second (2 cycle) fault clearing time.
3. For < 10 kA short circuit current available, the hazard/risk category required may be reduced by one number.
4. 65 kA short circuit current available, 0.33 second (20 cycle) fault clearing time.
5. 65 kA short circuit current available, up to 1.0 second (60 cycle) fault clearing time.
6. For < 25 kA short circuit current available, the hazard/risk category required may be reduced by one number.

33 *What happens if the FR rating of the PPE is less than the exposure?*

When a worker is exposed to incident energy that exceeds the rating of his or her protective clothing, an injury might occur. However, the worker's clothing will not ignite. The under-rated clothing will provide some thermal protection. Although an injury might occur, the arc-rated clothing will mitigate the exposure to some extent.

Will FR clothing protect me from shock or electrocution?

No. Electrical shock or electrocution is the result of electrical current flowing through a victim's body. Any time an exposed energized conductor is touched simultaneously with contact with earth, current will flow. To prevent electrical shock, the amount of current must not exceed 0.006 amperes. Only products that are tested to control current flow will do that. FR clothing is constructed from fabrics that may be conductive.

35 *Is fit important when wearing FR clothing?*

When the surface of the clothing is heated, the thermal energy is conducted through the fabric to the surface underneath. If the fabric is tight on a worker's skin, the skin could be burned by the energy conducted through the clothing.

A couple of principles can help guide you in seeking the right fit. First, movement should be unimpeded by a garment that is either too loose or too tight. Second, wearing layers may help protect from both flame and conduction. In general, you should choose your clothing based on the risks you are likely to face while doing particular tasks.

When is a face shield satisfactory face protection?

An arc-rated face shield worn in conjunction with an arc-rated balaclava (head sock) might provide adequate protection from arc flash. The face shield must be rated for both impact and thermal protection. Workers should recognize that the side of the face shield is open and might expose the side and back of the worker's head to thermal energy.

Why is it important to wear spectacles under a face shield or flash hood?

A face shield or a flash hood will protect your eyes from impact just as spectacles would. However, when you remove the face shield or the flash hood, your eyes are exposed just as if the spectacles have been removed (Figure 3). If

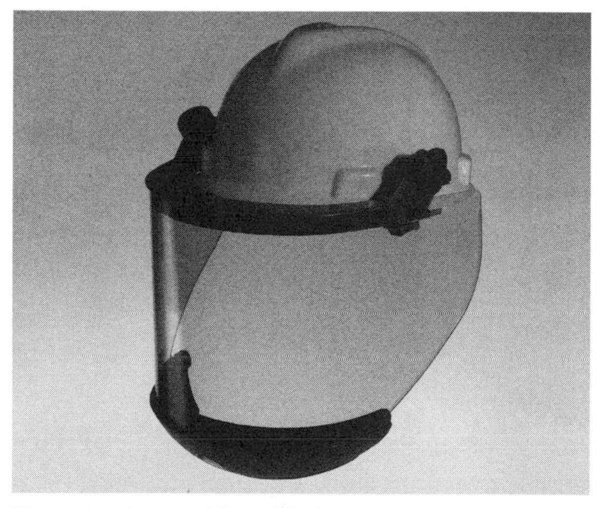

Figure 3 Arc-rated face shield.
Courtesy Paulson Manufacturing Corp.

you forget to put your spectacles back on, the eyes are unprotected for a longer period of time.

38 *What clothing must not be worn when exposure to arc flash exists?*

When an arc flash event occurs, the air (and gases that result from the plasma) will reach a temperature of 18,000°F to 20,000°F. Ordinary street clothing made from cotton or cotton blends will ignite at between 600°F and 1000°F. If a worker's clothing ignites, the worker's skin is exposed to the temperature of the burning clothing for several minutes. The result will be very significant burns. Clothing made from cotton, cotton blends, or any other flammable or meltable fabric must not be worn.

Any part of your body that is closer to a potential arcing fault than to the flash protection boundary must be protected. The PPE must have an arc rating at least equal to the estimated incident energy. The overall protective apparel may be assembled from several individual components, such as shirt and pants, provided the overall apparel covers all parts of the body exposed to the hazard. Note that a lab coat might protect the upper section of the person's body and not protect the lower portion of the body.

39 *What is the danger associated with the thermal hazard?*

The most significant result of the thermal hazard is the possibility of igniting a worker's clothing. Certainly a worker's unprotected skin is likely to be injured in the exposure. However, when a worker's clothing ignites, the duration of the exposure will be longer and result in a more severe injury.

40 *What is the best PPE practice for workers who may be exposed to arc flash?*

The most significant burn injuries occur when a worker's clothing ignites. In most cases, the duration of an arcing fault is short. Although the temperature of the arc plasma is very high, limiting the duration of the arc flash event limits the duration of the exposure. However, when a worker's clothing ignites, the worker's skin is exposed to the burning clothing for several minutes. When exposed to an arc flash event, workers tend to become disoriented and thus may have difficulty removing burning clothing, which further extends the exposure. This normally results in extensive burns.

Any arc-rated FR clothing will not ignite. All workers who are or may be exposed to an arc flash event should avoid clothing constructed from any non-FR fabric. Category 2 rated clothing has a "feel" that is similar to ordinary work clothing. Workers should wear arc-rated FR clothing that is rated as category 2 protection.

PART 5

Safety Grounds

When do I need to use temporary (safety) grounds?

What constitutes approval of temporary (safety) grounds?

More . . .

When do I need to use temporary (safety) grounds?

Temporary protective grounds sometimes are called safety grounds. Because the purpose of the temporary ground is to limit the voltage between adjacent points, the ground should be used whenever a chance exists that the deenergized conductor could become reenergized. For instance, any conductor that is outside a building is subject to re-energization by a lightning discharge.

Why must temporary (safety) grounds be approved?

When current flows in an electrical conductor, the interaction of magnetic forces results in physical force being applied to the conductor. When the amount of current is on the order of available fault current, the physical force on the conductor is significant. Temporary ground sets must be rated by the manufacturer and approved to handle the available fault current. The approval process enables the user to determine whether the temporary ground will meet the needs of the specific task or situation.

What constitutes approval of temporary (safety) grounds?

In general, manufacturers test and rate ground sets. No third-party approval is required. Workers should use only purchased ground sets (temporary or safety grounds). If the manufacturer rates the ground set, it is approved for applications that do not exceed the rating.

Approach Boundaries

What are approach boundaries?

What is the difference between shock boundaries and a flash protection boundary?

More . . .

44 *What are approach boundaries?*

Approach boundaries are imaginary three-dimensional figures that surround an exposed energized electrical conductor. The boundaries are measured in straight-line directions from each point on the conductor (Figure 4).

Why are approach boundaries important?

Dimensions of an approach boundary illustrate increasing exposure of a worker to an

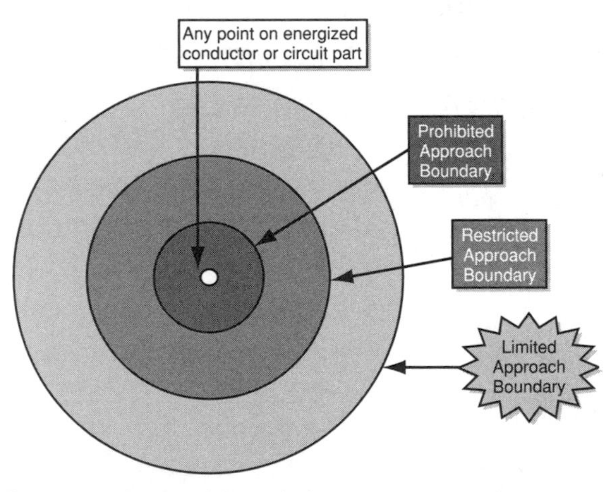

Figure 4 Approach boundaries.

exposed energized electrical conductor. As exposure to the conductor increases, the risk of injury also increases.

46 *What is the difference between shock boundaries and a flash protection boundary?*

Shock and electrocution are the result of current flowing through a victim's body. As defined by Ohm's Law, the amount of current depends on the voltage. Therefore, shock boundaries are voltage dependant.

The flash protection boundary defines the point where a person's skin might receive 1.2 calories of heat energy per square centimeter of surface area. Because the amount of incident energy depends on the capacity of the circuit to deliver thermal energy, the flash protection boundary depends on available energy.

Is arc flash boundary *a generally accepted term?*

Arc flash boundary is an undefined term that results in confusion; avoid using this term. The defined, and more appropriate term is *flash protection boundary*. The edge of an arc flash is not important from a personal safety perspective. The important characteristic is the distance where flash protective equipment is necessary to avoid injury.

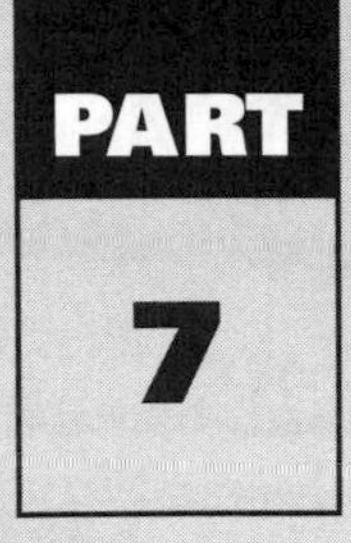

Safety Equipment

Is a GFCI effective in preventing injuries?

How are insulated tools rated?

More . . .

48 *Are safety interlocks important for personal protection?*

Generally, a safety interlock is associated with operating equipment or processes. Safety interlocks serve multiple purposes and are important for personal protection. They might be used to ensure that a manufacturing process is interrupted when the interlock operates. For instance, an overpressure switch or a high-level switch might be a safety interlock. A safety interlock might be a limit switch installed on a door to ensure that a segment of an operating process is interrupted when a door is opened.

Is a GFCI effective in preventing injuries?

Ground-fault circuit interrupters (GFCIs) are extremely effective in preventing injuries. These devices sense when current is flowing in an unintended circuit and assume that the current is through a person's body. A GFCI limits the amount of current to no more than 6 milliamperes, which is a level that will not result in electrocution (Figure 5).

Figure 5 Receptacle and circuit breaker configurations of ground-fault circuit interrupters (GFCIs).

50 *What is the role of overcurrent protection?*

Overcurrent protection is intended to remove the source of energy any time current exceeds the rating of the circuit. Overcurrent protection does not affect the shock or electrocution hazard. The speed at which an overcurrent device operates is one critical variable of available incident energy.

51 *Can I substitute a different size fuse?*

Like circuit breakers, fuses are tested and rated to clear a known level of overcurrent or fault. A worker may substitute fuses with smaller current ratings, provided the interrupting rating of the substitute fuse remains the same. However, larger fuses or fuses with smaller interrupting ratings must not be used. Fuses with the same ratings but from different manufacturers must not be mixed in the same circuit.

52 *Is electrical equipment tested in a failure mode?*

Electrical equipment that has a third-party label is tested and certified to comply with the requirements of a specific test procedure. However, most third-party testing does not consider arcing fault conditions. Unless the equipment is rated as arc resistant, arcing faults are not considered.

53 *Why is routine opening and closing of circuits an issue?*

When rated for the service, disconnecting means may be used to open or close a circuit. If the disconnecting means is not load rated, however, the handle should not be moved while the load is operating. Cable connectors should not be used to open or close a circuit, unless they are rated for the service.

When current is flowing in an electrical circuit, its inductive characteristic will not permit the current to be interrupted instantaneously. Each time a disconnecting means is operated, therefore, an arc occurs. Devices that are rated to open and close under load are designed to minimize the effects of the expected arc. However, the expected arc will damage devices that are not designed to open under load, increasing the chance of failure.

54 *When are insulated tools required?*

Some OSHA standards require insulated tools when working on or near an exposed energized electrical conductor (Figure 6). NFPA 70E has a similar requirement. Insulated tools are assigned a voltage rating. Insulated tools should be used for all work tasks in which a risk of contacting an exposed energized conductor exists, including any work task in which a dropped tool could contact an exposed energized conductor. However, insulated tools should not serve as the primary protection from shock or electrocution.

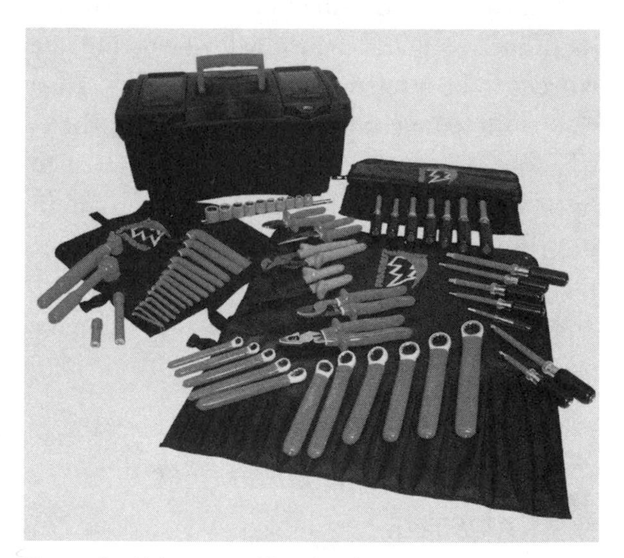

Figure 6 Voltage-rated hand tools.
Courtesy of Salisbury Electrical Safety, LLC

55 *How are insulated tools rated?*

Insulated and insulating tools are assigned a voltage rating by their manufacturer. Insulated hand tools are ordinary hand tools covered with a non-conductive material. For instance, socket sets are available that are rated at 1000 volts. Insulated tools that have a 1000-volt rating are marked with a double triangle (Figure 7).

Generally, tools intended for use on medium- or high-voltage circuits are constructed from insulating material. The physical length of the tool determines the rating of the insulating material. One example of an insulating tool is a live-line tool (hot-stick).

Figure 7 Double-triangle symbol for insulated tools.
Adapted with permission from F 1505-01 *Standard Specification for Insulated and Insulating Hand Tools*, © ASTM International, 100 Barr Harbor Drive, West Conshohocken, PA 19428.

Arc Flash/Arc Fault

What hazards are associated with arc flash?

What is the most frequent cause of an arcing fault?

More . . .

56 *What is arc flash?*

When a fault occurs in an electrical circuit (or a second fault occurs in an ungrounded circuit), an arc occurs. The arc is plasma that results from current flowing through air. The plasma converts some of the *electrical* energy into *thermal* energy at extremely high temperature. The temperature can approach 30,000°F. In most cases, the overcurrent device senses the fault and removes the source of energy, but the temperature frequently reaches 16,000°F to 18,000°F before the overcurrent device can clear the fault.

57 *What hazards are associated with arc flash?*

An arcing fault produces thermal energy and mechanical energy (a pressure wave). Some electrical energy is converted into visible light and other frequencies of the electromagnetic spectrum. Although the intensity of each of these forms of energy is not yet predictable, the thermal hazard is known to be severe, and the pressure wave is known to be significant.

In cooperation with NFPA, IEEE has initiated a multi-year research effort to determine all hazards and define a mechanism to predict their effects.

58 *Can an arc flash hazard exist in a manhole?*

Yes. Generally, cables and conductors in a manhole are high current or high voltage and serve several areas or devices. Although terminations normally do not exist in a manhole, splices are common, and terminations sometimes exist. The insulation on old cables can become brittle and subject to damage by moving the cable or conductor. The environment in the manhole is likely to be wet and congested.

Any work in a manhole subjects existing cables and splices to physical damage. In some cases, physical damage is likely to generate an arcing fault. Failures inside a manhole are infrequent; however, when a failure occurs, the elevated temperature created in the arcing fault surrounds any worker in the manhole.

59 *Are workers normally exposed to arc flash?*

Workers who operate disconnect switches might be exposed to injury should an arcing fault occur. However, if the installation meets the consensus requirements for overcurrent protection, the risk of injury is reduced. Equipment that is not adequately maintained increases the risk of injury. Some electrical equipment is constructed with ventilation holes in the door or cover. In those cases, the risk of exposure to injury is increased. Employers and workers must assess both the hazard and the risk of injury.

60 *What are the most important variables of an arcing fault?*

Three variables that primarily affect incident energy in an arcing fault are the capacity of the circuit to maintain the arc, the duration of the arc, and the distance between the worker and the arc. Employers should ensure that the overcurrent protective device is functional and operates in the minimum amount of time. A worker should position his or her body so that the distance between any body part and the potential arc source is as great as possible.

When is an arc fault most likely to occur?

Most arcing faults occur when something is moving. Opening a door, removing a cover, and operating a disconnecting means or a closing contactor are frequently actions that initiate an arcing fault. A worker's movement also might result in an arcing fault.

What is the most frequent cause of an arcing fault?

Noted psychologist H.W. Heinrich suggested that human errors are the primary cause of most incidents and injuries (H.W. Heinrich, D. Peterson, and N. Roos. *Industrial Accident Prevention: A Safety Management Approach.* Fifth Edition. McGraw-Hill, 1980). In the preamble to 29 CFR 1910 Subpart S, OSHA suggests that up to 67 percent of electrical injuries result from inappropriate action of a worker. The writer's experience concurs with these observations.

VOICES OF EXPERIENCE

"The fish tape grounded out to the live bussing and instantly vaporized."

A really good friend and workmate of mine was getting set up to pull some conductors into a 120/240-volt panel in a residential home (using EMT conduit). He had just received the end of the fish tape and was holding it out. He was having a conversation with his apprentice and, without thinking, he let go of the end of the fish tape. As it recoiled, the fish tape grounded out to the live bussing and instantly vaporized. He had turned towards the panel just as a fireball hit him in the face. Fortunately for his eyesight, he had his hands covering his face.

When it was all over, he had third-degree burns on his hands and face. Small metal fragments from the fish tape and bussing were embedded in his skin. He needed skin grafts to repair the burns. Proper safety gear would have saved his hands and face. Shutting down the power, or just simply paying attention when these hazards exist, would have prevented the accident altogether.

Doug Burrell
Product Development Engineer
Cheetah USA Corp
Sandy, Utah

The causes of all injuries and incidents can be divided into three categories:

- Unsafe equipment
- Unsafe conditions
- Unsafe action

Unsafe actions cause about two thirds of the total number of injuries and incidents; unsafe equipment and unsafe conditions combined cause the remainder of injuries and incidents. Although equipment does fail and workers are sometimes injured due to an unsafe condition, the action of a worker is the principal cause of an arcing fault.

63 *What work practice is most important to avoid injury from an arc flash?*

Workers are injured only when energy is released in an arcing fault or when they touch an exposed energized electrical conductor. Therefore, the only way to completely avoid the possibility of an injury from electrical energy is to remove the energy and take steps to ensure that the energy cannot reaccumulate. An arc flash event requires electrical energy. If no energy is available, no injury is possible. When that condition exists, the work task is considered to be in an electrically safe work condition. Therefore, the most important work practice is to create an electrically safe work condition. (See Question 89.)

64 *How does a worker know if he or she is exposed to a potential arc flash?*

In section 400.11, NFPA 70E suggests that a label be field installed on equipment that contains an arc flash hazard (Figure 8). In many

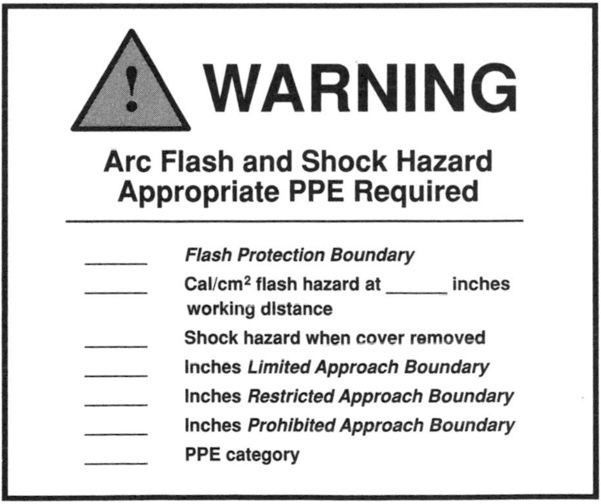

WARNING

Arc Flash and Shock Hazard
Appropriate PPE Required

_____ *Flash Protection Boundary*
_____ Cal/cm^2 flash hazard at _____ inches working distance
_____ Shock hazard when cover removed
_____ Inches *Limited Approach Boundary*
_____ Inches *Restricted Approach Boundary*
_____ Inches *Prohibited Approach Boundary*
_____ PPE category

Figure 8 Arc flash hazard label.

instances, the label identifies the arc rating of PPE that is necessary for protection from the estimated hazard. If a label is present, workers know that an arc flash hazard exists.

If a label is not present, the analysis must take a different form. The electrical safety program should contain a procedure that defines the necessary steps to perform the hazard/risk analysis. If neither a label nor a procedure exists, the worker must determine the capacity of the source of energy. If the capacity of the energy source exceeds 125 kVA, then you are exposed to a thermal hazard.

65 *What happens if the overcurrent protection fails to clear the fault?*

Circuit breakers and fuses may be applied improperly. Devices are sometimes installed in circuits so that available fault current exceeds the device's ability to clear a fault. If the overcurrent device is applied improperly and does not clear the fault, the device could fail violently and expel parts and pieces. The risk of injury is elevated significantly.

Why do I get different answers when I use different methods to calculate incident energy?

Each current method of predicting the thermal hazard associated with an arcing fault uses different variables and different mathematical processes. Each method of calculating incident energy has both positive and negative aspects. All methods produce a number that is an estimate, at best.

Am I exposed to an arc flash hazard when the equipment doors are closed?

An electrical installation that meets the requirements of the *National Electrical Code* does not expose a worker to arc flash hazard when all the code requirements are met. Other national consensus codes may provide the same protection. However, as an installation ages, the integrity of the installation deteriorates. A worker must consider the age and state of maintenance to help determine if exposure to arc flash might be present.

It is important to note that some electrical equipment includes ventilation holes in the door. An arcing fault in the enclosed equipment will direct the heated gases through the ventilation holes.

68 *Will rubber products protect me from arc flash?*

Rubber products are designed to resist the flow of current. They are neither designed nor intended to serve as protection from a thermal event. However, one characteristic of insulating rubber compounds is that although they will burn, they may be difficult to ignite. Because an arc flash event is likely to be very short, insulating rubber will provide substantial thermal insulation.

69 *How do I evaluate an arc flash hazard in a dc circuit?*

No consensus method exists to estimate incident energy associated with a dc energy source. Direct current flow is inherently different from alternating current flow, because an alternating current reaches zero twice each cycle and the arc is extinguished. The alternating current must re-ignite the plasma when the direction of the electron movement changes. A dc arcing fault might be more intense than an ac arcing fault. All current methods of estimating incident energy suggest that duration of the arc and the capacity of the energy source to provide current to the arc are the most important variables. It is likely that the same variables are most important in a dc arcing fault.

A dc fault could be evaluated as if it were an ac fault. The estimate might use dc circuit characteristics and apply them to current estimating methods for ac circuits. The resulting answer is unlikely to be accurate; however, the answer would provide a basis for selecting PPE. Any arc-rated PPE is an improvement over ordinary work clothing. DC circuits associated with batteries include hazards that are not present in ac circuits. If the dc circuit contains a significant component of pulsating current, the arcing fault is more likely to resemble an ac circuit fault.

70 *What is the difference between exposure to an ac circuit and a dc circuit?*

The electrical hazards are the same: shock or electrocution and arc flash. However, the degree or intensity of the hazard is likely to be different. No public information is available that provides information about tests conducted on dc arc flashes. An IEEE/NFPA collaborative effort is underway. The resulting research effort will provide guidance about estimating an arcing fault in a dc circuit.

Lockout/Tagout

Is tagout as safe as lockout?

What is the best practice for lockout/tagout?

More . . .

71 *What does one lock/ one person mean?*

A basic premise of lockout is that *each* person exposed or potentially exposed to a known hazard installs a lockout device on the source of the hazardous energy, resulting in one lock per person.

72 *Is it necessary to audit the lockout/tagout procedure?*

Yes. Employers can determine whether workers are implementing the lockout/tagout procedure or whether the procedure is effective only by auditing a lockout/tagout in progress. An effective audit provides information about both the procedure and the state of training. In every instance, the manner in which workers are expected to apply lockout/tagout to their work station, area, or task must be committed to writing and made available to all employees. Unless the procedure is in writing, each worker will have a different understanding of the requirements. Training on the procedure will also help to ensure that workers share an understanding of the requirements.

73 *What does in control of energy mean?*

As used when discussing lockout, the term *in control* means that a disconnecting means is physically restrained from operating. Normally, a lock is used to ensure that a worker is *in control* of potentially hazardous energy.

74 *Is tagout as safe as lockout?*

No. In transmission circuits, tagout is the only viable option, and transmission utilities implement effective alternatives. However, in all other instances lockout provides maximum safety.

75 *What is the best practice for lockout/tagout?*

Lockout is the best practice. Some employers tend to use tagout as the primary energy control system. They suggest that some equipment has no means to install a lock. Therefore, they revert to closing valves or opening disconnecting means and then install tags without locks. Tags can fall off and land on the floor, and they are sometimes misused. Some workers also place less importance on the tags' value. On the other hand, locks will stay in place. When properly installed, locks will prevent a valve or switch handle from moving.

The utility industry has used tagout successfully for many years. However, the key to the success seems to be related to the dedication of dispatchers. With due respect to the success related to transmission and distribution lines, installing both locks and tags is the best option.

76 *What is a person in charge?*

In most instances, when a lockout/tagout failure occurs and someone is injured, the root cause of the problem was ineffective or inadequate communication. It is therefore necessary to have a person in charge. When a single person is assigned the responsibility and held accountable to ensure that all sources of energy are and remain under control, the problem of inadequate or incomplete communication disappears.

77 *What is a simple lockout?*

A simple lockout is when equipment is supplied from a single source of energy and a single work crew is involved with the work task. Controlling the source of energy is simple: a single disconnecting means and a single lockout device.

78 *What is individual employee control?*

In some instances, a task must be performed within the enclosure that contains the disconnecting means. Frequently, any lock would swing away with the door. In other instances, the door cannot be opened with a lockout device installed. In these instances, the existence of a lock is moot. However, the worker must close and latch the door if he or she finds it necessary to leave the location for any reason. If the work task is not complete when the worker finds it necessary to leave the work location, he or she must employ simple lockout.

Checking for Absence of Voltage

How do I test for absence of voltage?

How do I choose a good voltmeter?

More . . .

79 What is the difference between measuring voltage and testing for absence of voltage?

Sometimes it is important to establish absence of voltage, and sometimes the intent is to determine the level of voltage. A worker needs to *measure* voltage to determine whether the voltage is 112 volts or 120 volts. When testing for absence of voltage, however, the only important issue is whether a voltage exists. Testing for absence of voltage is a go/no go or yes/no test.

80 How do I test for absence of voltage?

To ensure that both grounded and ungrounded circuits have no voltage available, a voltage test must include testing each phase conductor to every other phase conductor as well as testing each phase conductor to earth ground. The worker must ensure that the voltmeter rating is adequate for the circuit voltage. He or she must ensure that any adjustable controls on the meter are set to the appropriate scale also. The best practice for checking for absence of voltage is to use a single-function meter. If no adjustments are necessary and leads are hard-wired, the device cannot be set on the wrong scale.

Why is the voltmeter important?

Voltmeters are important because they are safety equipment. A voltmeter with an incorrect indication increases the risk of electrocution. If a voltmeter fails in service, an arcing fault can be the result. In fact, many injuries do occur because of voltmeter failures. Voltmeters are more important than some other safety devices because exposure to electrical injury exists in every instance. Therefore, you should keep voltmeters in a safe place and protect them from damage.

82 *How do I choose a good voltmeter?*

Voltmeters are designed and marketed for specific applications. Voltmeters purchased in electronics or home repair stores are likely to be designed for use in applications in which available energy is low. Such devices should not be used on industrial equipment and circuits where significant amounts of energy are available. Workers should ensure that each voltmeter used has a third-party label, complies with consensus standards, and has a Category 4 static discharge rating (Figure 9).

Figure 9 Multimeter.
Courtesy of Fluke Corporation. Reprinted with permission.

Definitions

What is the difference between a barrier and a barricade?

What is an electrically safe work condition?

More . . .

83 *What does anticipating failure mean?*

After electrical equipment is installed, it begins to deteriorate. Adequate maintenance extends the life of equipment. However, equipment could fail in service. Frequently, electrical equipment provides warning that failure is imminent. A worker might note an unusual noise or sound. An unusual smell frequently is a warning signal. Elevated temperature is a warning that a failure is near. Anticipating failure, then, is being cognizant of indications such as these and reacting to them.

When impending failure is recognized, remove the equipment load. Remove the source of energy with upstream disconnecting means. Do not operate disconnecting means that have an unusual smell, noise, or other indication of impending failure.

What is the difference between a barrier and a barricade?

A barrier is intended to provide a physical obstruction and prevent contact with an energized or potentially energized conductor. Barriers normally are used to isolate energized conductors. On the other hand, a barricade is a physical obstruction that is intended to serve as a warning.

What is blind reaching?

Workers sometimes attempt to locate or modify a circuit by reaching a hand into a location that is not directly visible, for instance, behind a barrier. Any location that is not in a direct line of sight is a blind location. The best practice is to avoid reaching into positions when visibility is not complete. Absence of adequate light effectively turns visible locations into blind locations. You should make certain that the intended contact point is adequately lighted and directly visible.

What is meant by the term degree of the hazard?

An incident that results from a hazard could be catastrophic. In slightly different conditions, another incident resulting from the same hazard might cause a small incident. The amount of energy available to the hazard determines the severity of an incident involving the hazard. For instance, each time a disconnecting means is closed into an energized circuit, an electrical arc results. When the arc is associated with a 15-ampere receptacle circuit, the arc is small. When the disconnecting means is in a 400-ampere circuit, the arc is larger. The degree of the hazard, then, refers to the potential severity of an incident resulting from the condition.

87 *What does duty cycle mean?*

Some electrical instruments are assigned a duty cycle that describes the maximum on time and the minimum off time. For instance, most solenoid-type voltage testers are assigned a duty cycle of 15 seconds. The period of time in an energized condition must not exceed the duty cycle. After using it energizes the instrument, the device must remain deenergized for at least the same length of time. The duty cycle defines the maximum time period that the device may be energized and the minimum time period that the device must rest before being

VOICES OF EXPERIENCE

"I know what I'm doing; I've done this type of work all my life!"

While in charge of a construction project to build a medium voltage substation, including installation of the medium voltage feeders and associated underground vaults, I was lucky enough to be assigned the very best journeyman electrician I have ever met. This doesn't mean that he is perfect when dealing with safety. He was in the process of switching supply feeders for the new substation, and was working in the electrical utility's vault. This is considered to be a confined space requiring PPE, an outside safety person, and an escape tripod and harness. This electrician had overlooked these safety precautions, and when the utility representative came around and found him working alone down in the vault in this manner, the electrician replied, "I know what I'm doing; I've done this type of work all my life!" The failure to follow safety procedures was brought to the safety inspector for the job, resulting in ordering the electrician to stop, get the safety equipment, and follow the procedures. He complied immediately.

When I talked to him about this later the same day, he exclaimed again, "I know what I'm doing; I've done this type of work all my life!"

The next morning, the electric utility's electricians were splicing onto the feeder that was disconnected the day before. This resulted in the feeder running across the roadway to another large vault with a 4-foot diameter manhole cover. When they energized the feeder after splicing was completed, the cables failed and arced. The resultant explosion and associated high temperatures and plasma blew the heavy manhole cover 15 feet into the air and cracked the concrete vault, destroying everything in the vault. My electrician was watching the process of energizing the feeder when it blew. I talked to him immediately after this occurred and he told me just how dumb it was to ignore the procedures, and that he now understood why these procedures exist.

I now have a story to relate to my contractors and electricians that emphasizes the importance of following the safety rules when working with medium voltage equipment. An extreme amount of energy can be released when a medium voltage circuit fails and arcs, and this incident emphasizes why we need to be so careful when working in vaults with energized conductors in them. Extreme care and planning must be observed when any medium voltage energized cable must be disturbed in any way. I just don't want to be anywhere in the vicinity when something like this blows up.

Arthur Warren, PE
Port of Seattle
Seattle, Washington

used again. The label on the voltmeter in Figure 10 features a duty cycle of 15 seconds.

88 *What are electrical safety program controls?*

Program controls are those policies that control the actions of each employee. For instance, a program might be based on an expectation that a circuit is considered energized until proven de-energized. A control might include an expectation that every employee is as-

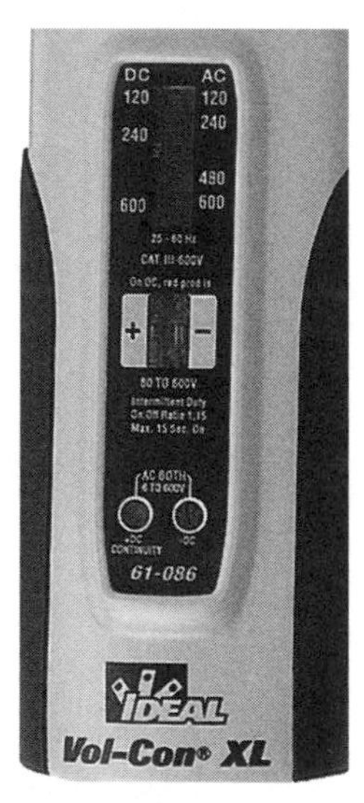

Figure 10 Label showing duty cycle.
Courtesy of Ideal Industries, Inc.

signed his or her own personal lockout devices. A control might also include an expectation that every circuit is energized until an electrically safe work condition exists.

89 *What is an electrically safe work condition?*

An electrically safe work condition refers to equipment and circuits in which all sources of electrical energy have been disconnected, verified as being open, and controlled by locks and tags. An electrically safe work condition exists only after the following six steps have been executed:

1. Determine all possible sources of electrical energy by checking up-to-date drawings, tags, and labels.
2. After properly removing the load current, open the disconnecting means.
3. Where possible, visually verify that an opening exists in all supply conductors.
4. Install locks and tags in accordance with an approved written procedure.
5. Verify the absence of voltage on all exposed conductors by checking the voltmeter for

proper operation both before and after verifying that no voltage exists.

6. Install safety grounds when it can be determined that conductors in the work area could become energized as the result of a static discharge, an overhead line falling, and similar circumstances.

90 *What is an energized work permit?*

Consensus and OSHA standards require employers to have an electrical safety program. The program must consist of procedures and policies that are necessary to provide a workplace that is free from recognized hazards.

One of these procedures must describe an energized work permit, including how to obtain and how to manage the process of generating and executing the permit. The employer's procedure must define responsibilities associated with the permit. The energized work permit ensures that work tasks that must be performed in the vicinity of an exposed energized electrical conductor receive sufficient attention. Workers making the choice to perform the task while the circuit is energized will have considered the increased potential of injury.

The written permit ensures that supervisory personnel know about the increased risk of injury. It also ensures that workers have the opportunity to say that the risk is unacceptable. When the energized work permit is complete and authorized, supervisors, managers, and workers all understand that the risk of injury to workers is elevated.

91 *Why is an energized work permit important?*

Managers and supervisors may not recognize that employees are executing a task with elevated risk of injury. An energized work permit provides an opportunity to educate managers (or customers) about the risks and hazards of a particular job. Requiring a supervisor's signature ensures that the supervisor understands that an increased risk of injury exists. The written permit provides the supervisor with the opportunity to delay the work until an electrically safe work condition exists. You can find an example of an energized work permit in Annex J of NFPA 70E.

92 *What is a hold card?*

In transmission circuits where the disconnecting means is located several miles from the point of work, a hold card is a device that is installed on the disconnecting means for a circuit on which work is progressing. A hold card is a method of communicating a warning that the disconnecting means must not be operated. The details of the hold card process vary among employers.

93 *What is incident energy?*

Incident energy is the thermal energy that contacts (is incident upon) a person's skin or clothing. Incident energy is defined in terms of heat energy per unit of area. The consensus designation is calories per square centimeter.

94 *What is a job briefing?*

A job briefing is a discussion of a work task before it is started. The briefing may be a short discussion of potential hazards, or it could be a complex discussion involving several workers or crafts. A job briefing may be known by different names in different organizations. An effective supervisor ensures that a job briefing is conducted at the beginning of each work task and at the beginning of each day.

Definitions

 What is a live part?

The term *live part* has been used for many years. However, electricians develop different understandings of the term's meaning. The term is defined in the *National Electrical Code* and NFPA 70E, and the committees who produce the standards are meticulous to ensure that the term is used in accordance with the definition. Other national consensus codes may not use the same definition, and workers do not necessarily apply the consensus definition.

Some workers consider that a part must be exposed for it to be a live part. Other workers consider any energized part to be a live part. In previous years, the presence of a hazard was necessary for a part to be considered a live part. The important issue is that a worker recognize if and when he or she is exposed to shock or electrocution. It may be helpful to clarify the understanding or meaning of this term before engaging in work tasks where the term will be used.

96 *What is a moveable conductor?*

When a worker approaches an exposed energized electrical conductor, the distance between the worker and the exposed energized conductor determines the risk of injury. In cases where the conductor is held in position mechanically and the worker is standing on a solid platform, the worker is in control of the approach distance; the conductor is fixed in position. However, in cases where the conductor moves with the wind or other external force, the worker is not in control of the approach distance; the conductor is moveable. When the worker is standing on a platform that moves, such as a platform on an articulating boom, the worker may not be in control of the approach distance. The conductor is considered to be a moveable conductor, because the distance between the worker and the conductor might vary beyond the worker's ability to control it.

97 *What is a procedure?*

A procedure is a document that describes how to perform a work task. For instance, a document containing a written step-by-step description of a process to install locks and tags on switches or valves is a procedure. A procedure might also describe how to perform a safety-oriented task, such as creating an electrically safe work condition, or it could describe how to check out new equipment.

What is a workplace?

A workplace is a location where employees are performing work. A workplace is associated neither with the discipline of the employee or the business nor with a product or service. If work is performed at a facility or location, the facility or location is a workplace.

What is a safe workplace?

The general duty clause of the OSH Act requires employers to provide a workplace that is free from recognized hazards. Although the clause suggests that employees must not be exposed to a known hazard, it does not mean that an injury cannot happen. The general duty clause means that employers must exercise normal care and take reasonable precautions to eliminate the risk of an injury. National consensus standards generally define normal and reasonable actions.

100 *What is a work practice?*

When workers execute a work task, each discrete step in the process is accomplished either by generating a new action or by repeating actions previously learned. The work might require a series of discrete steps or a few sets of steps. For instance, a worker might protect his or her voltmeter by storing it in a case. Another worker might keep his or her voltmeter loose in a toolbox. In one instance, the work practice is a good one and the other not so good. Using a ground-fault circuit interrupter on each cord is an example of a work practice.

Index

A

B

C

D

E

Index

F

G

H

M

N

O

P

R

S

T

Notes

Notes